THE ROMANES LECTURE
1898

Types of Scenery

and their

Influence on Literature

BY

SIR ARCHIBALD GEIKIE, D.C.L., F.R.S.

DELIVERED

IN THE SHELDONIAN THEATRE, JUNE 1, 1898

KENNIKAT PRESS
Port Washington, N. Y./London

TYPES OF SCENERY AND THEIR
INFLUENCE ON LITERATURE

First published in 1898
Reissued in 1970 by Kennikat Press
Library of Congress Catalog Card No: 72-113335
ISBN 0-8046-0954-3

Manufactured by Taylor Publishing Company   Dallas, Texas

TYPES OF SCENERY AND THEIR
INFLUENCE ON LITERATURE

TYPES OF SCENERY

AND

THEIR INFLUENCE ON LITERATURE

THE several races of mankind are marked off from each other by certain bodily and mental differences which, there can be no doubt, have been largely determined by the diverse geographical conditions of the surface of the globe. We may be disposed to put aside the first origin of these racial differences, as a problem which may for ever remain insoluble. Yet we cannot refuse to admit that in the disposition of land and sea, in the form and trend of coast-lines, in the grouping of mountains, valleys, and plains, in the disposition and flow of rivers, in the arrangement of climates, and in the distribution of vegetation and of animals, a series of influences must be recognized which have unquestionably played a large part in the successive stages of human development.

Though no record of the earliest of these stages has probably survived, some of the later steps in the progress of advancement may not be beyond the reach of investigation. The connexion, for example, between the circumscribing topography and geology of a country, and the mythological creed of its inhabitants offers a tempting field of inquiry, in which much may yet be done. Who can doubt that the legends and superstitions of ancient Greece took their form and colour in no small measure from the mingled climates, varied scenery, and rocky structure of that mountainous land, or that the grim titanic mythology of Scandinavia bears witness to its birth in a region of rugged snowy uplands, under gloomy and tempestuous skies?

If the primaeval efforts of the human imagination were thus stimulated by the more impressive features of the outer world, it is natural to believe that the same external influences would continue to exert their power during the later mental development of a people. In particular, it seems reasonable to anticipate that such potent causes would more or less make themselves felt in the growth of a national literature. The songs and ballads of the plains might be expected to present some marked diversities from those of the mountains. There may, of course, be risks of error in generalizations of this kind, especially where the writings of distinct races are compared with each other. But the risks may be reduced if we confine ourselves to the consideration of a single country and a single literature. Such at least is the task which I have undertaken on the present occasion. I propose to discuss the leading types of scenery that distinguish the British Isles, and to inquire how far it may be possible to trace from each of them an influence upon the growth of English literature.

Under the term scenery may be included those variations in the general aspect of the land that arise from the combined effects of three main geographical elements—topography, climate, and vegetation. Of these three factors, we are mainly concerned with the first, but the other two will necessarily obtrude themselves continually on our attention.

Although geological details are not essential for the inquiry before us, nevertheless some knowledge of them will be found of service in enabling us to recognize more clearly the essential features of a landscape, and to discriminate the real nature and extent of the diversities between the landscapes of different parts of the country. The fundamental elements of the scenery depend upon the nature and structure of the rocks that come to the surface, and the manifold varieties of external form arise from the constant succession of different geological formations.

The evolution of the scenery of these islands has been a long-continued process, wherein the chief part has been played by the air, rain, rivers, frost, and the other superficial agencies which are continually at work in wearing down the surface of the land. Each different kind of rock has yielded in its own fashion to the agents of destruction. Those which have offered least resistance have been worn down into hollows and plains. Those of a more durable nature have been left standing up as ridges, hills, or mountains. The composition of the rocks has likewise governed the nature of the soils and the distribution of vegetation. Chalk, for instance, forms long ridges of grassy and bushless down. Sandy and gravelly tracts are marked by heather and pine-trees. Volcanic rocks often rise into isolated crags with verdant slopes below. Granite lifts itself into craggy moorlands, with cliffs and corries and long trails of naked scree.

For the purposes of the present Lecture, I will arrange the scenery of this country in three leading types :—

I. LOWLANDS.

II. UPLANDS.

III. HIGHLANDS.

These types are often not separable from each other by any sharp lines of boundary. The lowlands, for example, include ranges of hills, and here and there gradually rise into uplands, which in turn occasionally mount into lofty, rugged ground that may well be called highland. Moreover, each type presents a number of local varieties, dependent on geological structure. Thus the English lowlands differ in many respects from the Scottish, and both from the Irish.

The arrangement now proposed, though not strictly scientific, is for the proposed discussion convenient. We shall find, I think, that each of the three main types has had a perceptible influence on our literature, and not only so, but that even the local variations of each type have left their impress on our literary history. To treat the subject as fully as it deserves to be treated would require a course of lectures. I can only attempt to illustrate it by selecting a few instances where the relation that I wish to establish seems to be most readily perceptible.

I. The Lowlands of Britain. If a line be drawn across England from the mouth of the Humber through the Midlands to the Bristol Channel, it will divide the country into two sharply contrasted portions. To the west of it, the older, harder, and therefore more durable rocks rise into the high grounds of Wales, the Pennine Chain, and the Lakes. To the east of it, on the other hand, the younger, softer, and consequently more destructible, formations occupy the whole of the territory, giving a characteristic lowland topography to the districts that have been longest settled and cultivated, which include the greater portion of what is most familiar and typical in English landscape.

The eastern Lowlands of England consist of a succession of gently undulating ridges, separated from each other by winding vales or plains. These features have a general trend across the country from southwest or west to north-east or east—a direction determined by the successive outcrops of the different formations. But the denudation of the surface has been so great and so unequal that, despite the continuity and parallelism of the formations, their outcrops have been worn into endless diversities of topographical detail, producing abundant variety among the gentle profiles of the landscapes.

As examples of the ridges, we may take the North and South Downs, and the chain of heights that stretches from Wiltshire into Norfolk. The intervening plains may be illustrated by the Weald of Sussex, and by the lower basin of the Thames.

The differences between the successive geological

formations in regard to composition and structure never pass a restricted limit. Essentially soft and easily decomposable, the rocks include no thick bands of much harder material than the rest, such as might project in prominent features. The prevailing characteristic of the topography is therefore unbroken placidity. Nowhere does any ruggedness obtrude itself. No part of the ground towers abruptly above or sinks suddenly beneath the general level. The successive ranges of low hills have rounded summits and smooth slopes, which, even when steep, seldom allow naked rock to appear, but conceal it everywhere under a carpet of herbage. So gentle is the inclination of the ground, from the watersheds to the coast, that the drainage winds in sluggish streams that often hardly seem to flow. ' Purling brooks ' are scarcely ever to be found. Since the running waters of the country have played no inconsiderable part in the inspiration of our poets, it is well worthy of remark that in the east and south-east of England the streams are, for the most part, silent. We shall see how strongly they are thus contrasted with the brooks that traverse the lowlands of the north.

Since the rocks of these eastern and southern districts decay into soils that are on the whole fertile, the surface is clothed with meadows, corn-fields, and gardens, while trees have been planted along the roads and fields, and in abundant patches of coppice and woodland. The country thus wears a verdant park-like aspect, which at once impresses the eye of a stranger who comes here either from the European continent or from America. Undoubtedly the progress of agriculture has gradually narrowed the area of the open heaths, while the ancient forests have wellnigh vanished. But there can be little doubt that the general enclosed and cultivated character of the landscapes has, over most of the ground, remained much the same as now for many hundred years.

One further feature of these English lowlands should be borne in mind. They are washed by the sea along the whole of their eastern and southern borders. Even the most inland part of them is not more than one hundred miles from the coast. Their long lines of ridge and down stretch to the very margin of the land, where they plunge in picturesque cliffs to the sea-level, as in the headlands of Flamborough, Dover, and Beachy. Moreover the coast-line is indented by numerous bays, creeks, and inlets, which, as they allow the sea to penetrate far into the land, furnish many admirable natural harbours. There can be no doubt that this feature in our topography has powerfully fostered that love of the sea which has always been a national characteristic, and has contributed to the development of that maritime power which has led to the establishment of our world-wide empire. To the same cause may be traced that appreciation of the poetry of the sea so noticeable in our literature.

We may now inquire how far the placid scenery of these eastern lowlands may have had an influence on the literary progress of the nation. It is of course chiefly among the poets that traces of such an influence may be expected to be discoverable. Not until comparatively late times did prose-writers deal with scenery,

save as a mere background for the human incidents
which they had to describe. It is impossible, within
the prescribed limits of a single lecture, to follow the
gradual development of an appreciation of landscape
among the writers who have lived on these English
lowlands. The simple child-like delight in Nature, so
characteristic of Chaucer, and the influence of cultivated scenery, so conspicuous in him, are readily
traceable among his successors. Shakespeare throughout his plays presents us with not a few reminiscences
of his youth among the Warwickshire woodlands. In
Milton we see how the placid rural quiet of the Colne
valley inspired the two finest lyrics in the English
tongue. For a century after Milton's time, poetry in
this country ceased to have any living hold on outer
nature, but became with each generation more polished
and artificial. When at last a reaction set in, the impulse
that led to the most momentous revolution that has
marked the history of English poetry came in large
measure from the writings of three poets, each of whom
drew his inspiration from lowland scenery. As illustrations therefore of the part played by this scenery in our
literature, I will cite Cowper, Thomson, and Burns.

The retreat in the valley of the Ouse, to which
Cowper escaped from the noise and distractions of city-
life, was eminently fitted by its quiet seclusion to soothe
his spirit, and to fill his eye and his imagination with
images of rural peacefulness and gentle beauty. If the
choice of such a home was of infinite service to the
poet, it was hardly less momentous in the progress of
English literature.

The scenery of that valley, around Olney and Weston, is thoroughly characteristic of the southern lowlands. The ground lies on limestones and clays, belonging to the Oolitic series, which, though they have been greatly denuded, have yielded in a general equable manner to the progress of decay. They possess no such differences of structure as to allow one part of them to project in crag or scar above the rest. They have been worn down into a gently undulating plateau or plain, covered with corn-fields and pastures, and dotted with occasional woods and 'spinnies,' or patches of trees. Farms and villages diversify the landscape, while to the north lies the forest-like expanse of Yardley Chase. Through this champaign the River Ouse has cut for itself a winding valley, the bottom of which, quite flat and from a few hundred yards to upwards of a mile in breadth, lies rather more than a hundred feet below the general level of the country. Along the flat alluvial plain, the stream, sluggishly flowing among rushes and sedges, curves in circuitous bends, often dividing so as to enclose insular patches of meadow, which it entirely overspreads in times of flood. The slopes rise softly from the river-plain, now projecting now retreating, as the valley winds from side to side. A gentle ascent brings us from the banks of the Ouse up to the highest part of the ground in the vicinity, and places before our eyes a wide sweep of rich agricultural country, with peeps of village spires and gleams of the winding river.

Such were the simple elements of Cowper's landscape. They have no special attraction that is not

shared by hundreds of other similar scenes in the
Oolitic tracts of England. To the cursory visitor they
may even seem tame and commonplace. And yet for
us, apart from any mere beauty they may possess, they
have been for ever glorified and consecrated by the
imagination of the poet. We see in them the natural
features which soothed his sorrow and gladdened his
heart, and which became the sources of an inspiration
that breathed fresh life into the poetry of England. The
lapse of time has left the scene essentially unchanged.
We may take the same walks that Cowper loved, and
see the same prospects that charmed his eyes and filled
his verse. In so following his steps, we note the accu-
racy and felicity of his descriptions, and appreciate
more vividly the poetic genius which, out of such
simple materials, could work such a permanent change
in the attitude of his countrymen towards nature. As
an illustration of his treatment of landscape let me
cite the well-known passage descriptive of the walk
between Olney and Weston—

> How oft upon yon eminence our pace
> Has slackened to a pause, and we have borne
> The ruffling wind, scarce conscious that it blew,
> While admiration, feeding at the eye,
> And still unsated, dwelt upon the scene.
> Thence with what pleasure have we just discerned
> The distant plough slow-moving, and beside
> His labouring team, that swerved not from the track,
> The sturdy swain diminished to a boy!
> Here Ouse, slow-winding through a level plain
> Of spacious meads with cattle sprinkled o'er,
> Conducts the eye along his sinuous course
> Delighted. There, fast rooted in their bank,

> Stand, never overlooked, our favourite elms,
> That screen the huntsman's solitary hut;
> While far beyond, and overthwart the stream,
> That, as with molten glass, inlays the vale,
> The sloping land recedes into the clouds;
> Displaying, on its varied side, the grace
> Of hedge-row beauties numberless, square tower,
> Tall spire, from which the sound of cheerful bells
> Just undulates upon the listening ear,
> Groves, heaths, and smoking villages remote[1].

No scene could have been more thoroughly congenial to such a temperament as that of Cowper. He never wearied of the sights and sounds of that peaceful landscape. He watched its changes from hour to hour, from day to day, and from season to season. Every change awakened new joy in his breast, and gave fresh inspiration to his verse. And so year after year he lived in closest communion with nature. Well might he say that

> Scenes must be beautiful which, daily viewed
> Please daily, and whose novelty survives
> Long knowledge and the scrutiny of years[2].

Cowper's poetic vision was like his landscape, limited, though within its range it was searching and accurate. His timid nature shrank from what was rugged and wild. He found his consolation in

> Nature in her cultivated trim
> And dressed to his taste[3].

But no one before him had revealed to men the infinite variety and beauty and charm to be seen by the

---

[1] 'The Task,' bk. i. 154–176.
[2] Ibid. i. 177.   [3] Ibid. iii. 357.

contemplative eye even in these everyday surroundings. The calm of evening—

> With matron step slow moving, while the Night
> Treads on her sweeping train [1],

or the river shining in the moonlight beneath the 'wearisome but needful length' of Olney bridge, or the creep of autumn over wood and field and weedy fallow, or the advent of winter and the shrouding of the valley under snow, or the coming of spring, when—

> The primrose ere her time
> Peeps through the moss that clothes the hawthorn root [2],

—every mood of nature in his sequestered vale is painted with a vividness and skill that evoke our admiration, and with a sympathy and grace which win our heart. That quiet valley has thus become classic ground, for as long as English poetry is read, the affections of men will be drawn to the home of Cowper by the banks of the Ouse.

The other two lowland writers whom I have selected were contemporaries of Cowper, though Thomson died when Cowper was only seventeen years old. In dealing with their influence, we turn to the Scottish lowlands where they both were born, and from which came their poetic impulse.

A traveller, familiar with the low grounds of England, when he first enters the northern lowlands is at once impressed by their much more limited extent. He finds them to occupy comparatively restricted spaces between uplands and highlands, their largest expanse lying in the

---

[1] 'The Task,' iv. 246. [2] Ibid. vi. 112.

broad depression between the southern border of the Highland mountains and the great tract of high pastoral land in the southern counties. Another considerable area of them intervenes between these southern uplands and the flanks of the Cheviot Hills. A further contrast to the English type is to be seen in the broken character of the surface. There are no extensive level flats like those of the English midlands, nor any continuous bands of gentle ridge like those of the English downs. The ground is separated into detached districts by ranges of hills, which sometimes even become sufficiently high and broad to deserve the name of uplands. A third characteristic arises from the great diversity of geological structure, and especially the intercalation of abundant volcanic rocks among the other formations. As a consequence of this intermingling of hard and durable materials with others more easily abraded, the topography is diversified by endless crags and hills rising picturesquely out of the surrounding lower country and often crowned with ancient castles or ruined peels.

Such a disposition of the elements of the landscape is accompanied with a much greater diversity in the mantle of vegetation than is seen among the English lowlands. The surrounding uplands and hills, for the most part bare of trees, are clothed with pasture, save where they support a covering of heathy herbage or dark peat-moss. The lower grounds have in large measure been brought under cultivation, but still retain tracts of moorland, haunted by lapwing and curlew. Though the ancient natural wood has disappeared,

hard-wood trees are now abundant, while sombre
plantations of fir give a northern character to the
landscapes.

But perhaps the feature in these Scottish lowlands
which more particularly deserves notice here, is the
contrast to be found between their streams and those of
south-eastern England. Owing to the uneven form and
steeper slope of the ground, the drainage runs off rapidly
to the sea. The brooks are full of motion, as they tumble
over waterfalls, plunge through rocky ravines, and sweep
round the boulders that cumber their channels. They
furnish, moreover, countless dells and dingles where
the native copsewoods find their surest shelter. There
the gorse and the sloe come earliest into bloom, and
the wild flowers linger longest. There too the birds
make their chief home. These strips of wild nature,
winding through cultivated field or bare moor, from
the hills to the sea, offer in summer scenes of perfect
repose. But they furnish too, from time to time,
pictures of tumult and uproar, when rain-clouds have
burst upon the uplands, and the streams come down
in heavy flood, pouring through the glens with a din
that can be heard from far.

Brooks and rivers have always had a fascination for
poets. Their banks supply secluded spots for reverie
and communion with nature. Under their shade and
shelter, bird and blossom and flower are guarded from
the blasts around, while the sparkle and murmur of
their waters give a sense of life and companionship
even in the depths of solitude. Constant familiarity
with such a type of stream as that of the Scottish

lowlands could hardly fail to strike the imagination in a different way from that which has attended the slow-creeping and silent brooks of the south-east of England. The Scottish poets, even in the earlier centuries, show traces of the influence of their more rugged surroundings; but not till the last century did this influence manifest itself in such a manner as to affect the general flow of English literature.

Of the two Scottish lowland poets whom we have now to notice, James Thomson was considerably the earlier. He was born in Roxburghshire in 1700, within hearing of the ripple of the Tweed, within sight of the Cheviot and Lammermuir Hills, and in a region famous in Border ballad and song. To the east the uplands of the Cheviot Hills rise as a blue ridge, high enough to come often within the clouds, to catch the first snows of autumn, and to keep them unmelted in northern rifts until the spring. From these long, bare undulating uplands a number of streams descend northwards into the Tweed, each having its own dale, with its own ridge of moors on either hand, and its meadows and cornfields along the bottom. The slope of the ground gives these descending waters such an impetus as sends them dashing over rocky channels, here and there cutting a scaur or ravine, murmuring over gravelly bottoms, winding through flat haughs, and finally finding their way into the Tweed. The water-courses are thus in themselves full of variety and life, and their charms are enhanced by the alternation of meadow and field, coppice, ferny brake and woodland, through which they wander. Nor are occasional bolder features wanting to enliven these quiet valleys. Here and there knobs of volcanic material rise along the crests of the ridges into prominent hills, which are conspicuous landmarks all over the Border country. Since Thomson's day the plough has no doubt crept further up the hill sides; more wood has been planted and more ground has been enclosed. But there is still plenty of bare moor and peat-moss, and the pastoral character of the district yet remains. The traditions of Border warfare have grown fainter, but the ruined peel-towers still stand as picturesque relics of the old wild times. The climate, too, has not changed. The winter storms still send down the rivers in full flood, and bury the vales in deep snow; the spring whitens the meadows and hedgerows with flower and blossom, and the short summer gives way to an early autumn.

Such was the scenery that inspired the 'sweet poet of the year,' as Burns called him. Thomson came to London in 1725, when he was twenty-five years of age, and the following year he published his 'Winter.' The poem, though written at Barnet, took its inspiration from the Border. The verse was turgid, full of latinisms, and sadly lacking in the simplicity and directness which the subject required. Nevertheless it could not conceal the genuine poetic gift of the writer, and it immediately became popular, for notwithstanding its artificial style, it took the reader at once into the sanctuaries of nature, from which the poetry of the previous hundred years had been exiled. It awakened a general interest in features of landscape never before described so fully or so well in English verse. It painted changes of the sky—tempest, rain-clouds, and snow-storms, and it brought the gloom of a northern winter vividly before the imagination of dwellers in a more southerly clime.

Thomson told the world how in his youth,

> Nurṣed by careless solitude he lived
> And sang of nature with unceasing joy[1],

and how, with 'nature's volume broad displayed,' it was his sole delight to read therein, happy if it might be his good fortune,

> Catching inspiration thence
> Some easy passage, raptured, to translate[2].

He had been used in his early years to muse

> On rocks and hills and towers and wandering streams[3],

and these now became the subjects of his song. Thomson, like his greater successor Burns, had from earliest boyhood been familiar with the burns and waters of his northern home. When he came to England he found but little entertainment in the landscapes around London, and longed for 'the living stream, the airy mountain, and the hanging rock.' He portrays with evident delight the changeful aspect of his native watercourses in the various seasons of the year. He knew well the 'deep morass' and 'shaking wilderness,' where many of them 'rise high among the hills,' and whence they assumed their 'mossy-tinctured' hue. He traces them as they 'roll o'er their rocky channel' until

---

[1] 'Winter,' 8.   [2] 'Summer,' 192.   [3] Ibid. 89.

they at last lose themselves in 'the ample river' Tweed[1].
He describes them as they appear at sheep-washing
time, and dwells on their delights for boys as bathing-
places. But it is their wilder moods that dwell most
vividly in his memory, when

> From the hills
> O'er rocks and woods, in broad brown cataracts,
> A thousand snow-fed torrents shoot at once[2].

It is worthy of remark, however, that even though
nature is his theme, the poet writes rather as an in-
terested spectator than as an earnest votary. He reveals
no passion for the landscapes he depicts. He never
appears as if himself a portion of the scene, alive with
sympathy in all the varying moods of nature. His verse
has no flashes of inspiration, such as contact with storm
and spate drew from Burns. It was already, however,
a great achievement that Thomson broke through the
conventionalities of the time, and led his countrymen
once more to the green fields, the moors, and the
woodlands.

In the successive poems which when placed together
made the 'Seasons,' published in 1730, Thomson con-
tinued to draw on his recollections of the Scottish
Border for the descriptions of landscape that form so
fundamental a part of his theme. It is interesting to
note that even after he had been five years in the south
of England, and must have seen in that time much
variety of weather and many different watercourses,
it is still from the north that he draws his sketches.
When, for instance, he tells how in autumn,

---

[1] 'Autumn,' 476.   [2] 'Spring,' 381, 400, 402; 'Summer,' 13.

> Red from the hills, innumerable streams
> Tumultuous roar[1],

the colour of his torrents betrays their Scottish origin.
He was thinking of the spates in his native streams
which sweep across tracts of Old Red Sandstone, and
come down almost brick-red in hue.   There are no red
rocks, and therefore no red brooks in Middlesex and
the surrounding districts.

But before the completion of the 'Seasons' the influence of English lowland scenery had begun to impress
itself on Thomson's imagination.   The softer and
ampler landscape of fruitful plains, with its richer agriculture and fuller population, its farms, villages, and
country houses, filled his mind with a new pleasure.
Some trace of this widened experience may be seen
in the additions successively made to the earlier
poems, such as the picture of Hagley Park introduced
into the poem on 'Spring.' But it was in his last
effusion, 'The Castle of Indolence,' that this English
influence gained entire sway, and the Scottish memories
faded into the background. Here we find ourselves
amid the typical landscapes of the south of England
—landscapes, however, so transfused by poetic genius
as to acquire an individuality of their own. We are
led into 'a lowly dale fast by a river's side'; we wander
through 'sleep-soothing groves and quiet lawns between'; we see 'glittering streamlets' in a sunny glade ;
we skirt a 'sable, silent, solemn forest'; and pass a
'wood of blackening pines,' which runs up the hills on

---

[1] 'Autumn,' 337.

either side, and see the famous castle 'close hid amid embowering trees,' that make a kind of chequered day and night. The landscape, we are told, 'inspires perfect ease'—a quality which must now have become indispensable to the 'bard, more fat than bard beseems.' Thomson, with his adoption of English scenery, had also polished his style and rid himself of much of his turgidity and latinism. He had changed his theme, too, and had chosen one more in consonance with the prevailing vogue. But to the end he had an eye for the charms of the free open face of nature. For the share he took in bringing back into our literature the recognition of these charms, he will ever hold an honourable place in the history of letters.

It was from another and somewhat dissimilar part of the Scottish lowlands that a far more powerful impulse than that of Thomson was given by the genius of Burns to the progress of the literary revolution of last century. The landscapes of Ayrshire, where Burns was born and spent most of his life, though akin in their main aspects to those of Roxburghshire, present nevertheless certain well-marked differences in topography which were not without their influence on the muse that inspired 'Tam o' Shanter' and 'Halloween.'

The lowlands familiar to Burns throughout most of his life form a wide undulating plain, surrounded on three sides by ranges of upland, and on the fourth by the open Firth of Clyde. The heights along the southern side belong to the long and broad chain of uplands which stretches from Portpatrick to Saint Abb's Head. Reaching heights of sometimes more than 2,500 feet above the sea, they stretch as a wide pastoral country, much of which is still covered with 'muirs and mosses many.' These high grounds catch the clouds and mists from the Atlantic, and receive such a copious rainfall as to feed many large streams which cross the lowlands to the sea. The number and size of these streams form a notable feature in the scenery, and the different geological formations through which they flow have contributed to give much variety to their channels. Here they may be seen flowing in a narrow glen, there opening into a wider strath, or creeping sullenly in a narrow chasm between precipitous walls of naked stone, or dashing merrily over rock and boulder beneath overarching trees, or sweeping in wide curves through open meadows or dense woods, and finally carrying their burden of mossy water into the blue firth.

These streams, with their endless changes of aspect, their variations from season to season, their play of sunshine and shadow, their wild flowers and their birds, had a strong hold on the affections of Robert Burns. His best inspiration came to him from them. As he tells us himself—

> The Muse, na Poet ever fand her,
> Till by himsel he learn'd to wander,
> Adown some trottin' burn's meander,
>   An' no think lang;
> O sweet to stray, an' pensive ponder
>   A heart-felt sang[1].

In the poem from which these lines are quoted, after

---

[1] 'To William Simpson,' stanza 15.

alluding to the poetic fame of other streams, while those of his own county remained unsung, the poet declares his resolve to atone for this neglect.

> We'll gar our streams and burnies shine
>   Up wi' the best.
> We'll sing auld Coila's plains an' fells,
> Her moors red-brown wi' heather-bells,
> Her banks an' braes, her dens an' dells.

Amply did he fulfil his promise. There is not a river, hardly even a tributary, within his reach, that has not been made famous in his lyrics. In the first bloom of opening manhood it was the Ayr and the Doon that gave him inspiration, and when broken in health and spirits, and with an early grave opening before him, it was by the banks of the Nith that his last poetic impulse arose.

In his relation to Nature there was this great difference between Burns and his literary contemporaries and immediate predecessors, that whereas even the best of them wrote rather as pleased spectators of the country, with all its infinite variety of form and colour, of life and sound, of calm and storm, he sang as one into whose very inmost heart the power of these things had entered. For the first time in English literature the burning ardour of a passionate soul went out in tumultuous joy towards Nature. The hills and woods, the streams and dells were to Burns not merely enjoyable scenes to be visited and described. They became part of his very being. In their changeful aspects he found the counterpart of his own variable moods; they ministered to his joys, they soothed his sorrows. They yielded him a companionship that never palled, a sympathy that never failed. They kindled his poetic ardour, and became themselves the subjects of his song. He loved them with all the overpowering intensity of his affectionate nature, and his feelings found vent in an exuberance of appreciation which had never before been heard in verse.

Among the natural objects which exerted this potent sway over the poetry of Burns, the streams of Ayrshire and Nithsdale ever held a foremost place. Their banks were his favourite haunt for reverie. They were familiar to him under every change of sky and season, from firth to fell. Each feature in their seaward course was noted by his quick eye, and treasured in his loving memory. Their union of ruggedness and verdure, of sombre woods and open haughs, of dark cliff and bright meadow, of brawling current and stealthy flow, furnished that variety which captivated his fancy, and found such fitting transposition to his verse. His descriptions and allusions, however, are never laboured and prominent; they are dashed off with the careless ease of a master-artist, whose main theme is the portrayal of human feeling. Even when the banks and braes have been the immediate source of his inspiration, Burns quickly passes from them into the world of emotion to which he makes them subservient.

So numerous and descriptive are his allusions to them that a luminous account of the characteristics of the Carrick brooks and rivers might easily be compiled from Burns' poems. At one moment we find him appealing to them for sympathy in his grief:

> Ye hazelly shaws and briery dens,
> Ye burnies, wimplin' down your glens
>   Wi' toddlin' din,
> Or foaming, strang, wi' hasty stens,
>   Frae linn to linn [1].

The same appeal forms the burden of his song on the 'Banks and braes o' bonnie Doon.' He leads us where

> In gowany glens the burnie strays,
> Or trots by hazelly shaws and braes,
>   Wi' hawthorns gray [2].

He pictures the stream after a rain-storm, when

> Tumbling brown, the burn comes down,
> And roars frae bank to brae [3];

or when the breath of the Atlantic has swept over the wintry hills and the

> Burns wi' snawy wreeths up-choked
>   Wild-eddying swirl,
> Or through the mining outlet bocked
>   Down-headlong hurl [4].

But nowhere does his delight in these features of his native landscape find more exuberant expression than in his 'Halloween,' when he interrupts his narrative of Leezie's misadventure to give a graphic picture of one of his brooks in the calm moonlight of an autumn evening.

> Whyles owre a linn the burnie plays,
>   As thro' the glen it wimpl't;
> Whyles round a rocky scaur it strays
>   Whyles in a wiel it dimpl't;

---

[1] 'Elegy on Captain Matthew Henderson.'
[2] 'Poem on Pastoral Poetry.'   [3] 'Winter, a Dirge.'
[4] 'A Winter Night.'

Whyles glitter'd to the nightly rays,
·Wi' bickerin', dancin' dazzle;
Whyles cookit underneath the braes,
Below the spreading hazel
Unseen that night.

Born near the River Ayr, and having spent his boyhood and youth in its valley, Burns had ever a special
affection for that stream, along whose banks he had
composed some of his finest poems. When his enforced
emigration to America was settled, and his trunk on its
way to the ship, he wrote a parting song in the burden of
which the banks of Ayr are made to stand for his native
country as a whole :—

The bursting tears my heart declare,
Farewell, the bonnie banks of Ayr[1].

When the respite came, and he found himself famous
and in Edinburgh, the Address which he wrote to the
Scottish capital contrasted his reception there with what
had gone before, and again his heart was by his beloved
river :—

From marking wildly-scattered flowers,
As on the banks of Ayr I stray'd,
And singing, lone, the ling'ring hours,
I shelter in thy honour'd shade.

And lastly, when the shadows were beginning to
gather around him at Ellisland, his thoughts would
go back to the same scene. In one of his latest and
most pathetic songs we find once more a reminiscence
of his associations with the river of his youth :—

---

[1] 'The gloomy night is gathering fast.'

Ayr gurgling kiss'd his pebbled shore
  O'erhung with wild woods, thick'ning green ;
The fragrant birch, and hawthorn hoar,
  Twin'd am'rous round the raptur'd scene.
The flowers sprang wanton to be prest,
  The birds sang love on every spray,
Till too, too soon, the glowing west,
  Proclaim'd the speed of winged day[1].

When Burns moved from Ayrshire to Nithsdale, he found at his new home another valley and another river that could minister to his inspiration. The Nith took the place of the Ayr. But it could not wholly fill that place, for its landscape is less ample, the hills come closer down upon the valley, while the river, in its lower course, curves from side to side in a wide alluvial plain, without the variety that marks the lower part of the Ayr. We seem to recognize the influence of these differences in the allusions in the songs.

The landscapes of Burns are marked by some curious limitations. Though he was born within sight of the picturesque mountain group of Arran, it does not come within his poetic outlook[2]. Though the ' craggy ocean pyramid' of the Clyde rose so stupendously from the firth in front of him, he makes no use of it further than to tell how ' Meg was deaf as Ailsa Craig.' Its distant grandeur does not seem to have struck his imagination. Indeed, if we examine his treatment of scenery, we may observe that it is the nearer detail that appeals to him. His pictures are exquisite foregrounds with seldom any

---

[1] 'To Mary in Heaven.'

[2] This was remarked by Wordsworth in the prefatory note to his lines on Mossgiel.

distinct distance.   But perhaps more remarkable still is the small place which the sea takes in the poetry of Burns. We must bear in mind that he was born and spent his boyhood within sight and hearing of the open Firth of Clyde.   The dash of the breakers along the sandy beach behind his father's 'clay biggin' must have been one of the most familiar sounds to his young ears.   Yet the allusions to the sea in his poems betray little trace of this association.   They are in large measure introduced to mark the wide distance between separated friends.

It must be remembered, however, that his life, after he began to write, was passed inland, where the wide firth could only be seen from the rising ground at a distance of several miles.   Yet Burns has left testimony that his imagination had not been insensible to the life and movement of the ocean.   One of the most effective touches in his picture of the night scene in the 'Brigs of Ayr' is given in the reference to the neighbouring sea—

> The tide-swollen Firth, wi' sullen-sounding roar,
> Through the still night dashed hoarse along the shore;

and when his native Muse gives him her benediction she tells how she had watched his passionate love of Nature :—

> I saw thee seek the sounding shore,
> Delighted with the dashing roar;
> Or when the North his fleecy store
>   Drove thro' the sky;
> I saw grim Nature's visage hoar
>   Strike thy young eye[1].

---

[1] 'The Vision.'

II. The UPLANDS of the British Isles consist of undulating plains or plateaux which lie from 1,000 to more than 2,000 feet above the sea. Seen from a distance, they look like ranges of hill or mountain, but without that variety of peak and crest which a true mountain outline would present. Though they may rise steeply out of the lower grounds, we have only to climb to their summit to find ourselves at the edge of a wide rolling platform, which may stretch for leagues without ever rising into any sharp prominence, or departing from the same monotony of moorland. Yet if we attempt to cross this seemingly continuous tableland, we find our progress barred by many valleys which, deep sunk beneath the general level, divide the plateau into separate blocks or ridges.

The surface of these uplands is for the most part treeless and even bushless. Where not covered with peat-moss, it is clothed with bent or with heather, kept short and green by periodical burning in the spring-time. Herds of cattle and flocks of sheep wander over the pastures, but there may be little other visible trace of human occupation upon these heights. It is in the little hollows that lead down into the main valleys, and in these valleys themselves, that trees make their appearance, first in scattered saplings of birch or mountain-ash, and then in thicker copsewoods or in artificial plantations of fir and larch. In these sheltered depressions, the farms and villages of the region have been planted, and cultivation has been slowly pushed upward on the slopes of the fells. Thus the larger part of the area of the uplands is uninhabited, the population being restricted to the more or less sheltered 'hopes,' hollows, dales, and valleys.

This type of scenery presents many local varieties, according to the geological structure of the ground. Where the rocks have been but little disturbed, the sides of the valleys display a succession of parallel bars of stone with intervening grassy slopes, such as may be seen among the moors of the East Riding, or in the dales of the Pennine Chain. Where, on the other hand, the rocks have been much compressed and pushed over each other by powerful movements of the terrestrial crust, their erosion has given rise to no regular topography, but they decay into rounded outlines and are covered over with heath and herbage, as in South Wales and southern Scotland.

Of the British uplands, the only district that claims notice here in connexion with our literature is that of the wide Border country of England and Scotland. It stretches through the moorlands of Northumberland and Cumberland into the range of the Cheviots on the one hand, and on the other into the great tract of high ground, which extends through the Lammermuir and other groups of fells from the North Sea to the Solway Firth. For many centuries this region has been pre-eminently pastoral. The natural forest, which in old times clothed much of its surface, has almost wholly disappeared before modern agriculture, and the plough has in successive generations crept higher up the slopes from the meadows of the dales. But there can be little doubt that though roads and railways have done much to open up these solitudes, the natural features remain essentially unchanged.

It was among these uplands that the Border ballads had their birth. We may therefore pause for a few moments to inquire what trace may still be discernible of the influence of the landscape upon the tales of war and love, of feud and raid and rescue, which have made that Border-land famous in our literature.

At the outset it is desirable to realize the all-important character of the valleys in the human history of the uplands. From time immemorial, these strips of more sheltered and cultivable ground, deep sunk below the general level of the moorlands, have been to a large extent cut off from each other by high tracts of fell and moss. Each of them took its name from the stream which, rising far up among the moors, and gathering tributary rivulets from glens on either side, winds down the strip of haugh along the valley-bottom. For generations past the people have looked on their native stream with an affectionate regard[1]. It has been the bond of union that has made the natives of each dale into one family or brotherhood. The valley itself may vary its scenery as it passes across different parts of the upland, here narrowing into a glen, there widening into a strath ;

---

[1] Scott was familiar with this natural trait. ' " That's the Forth," said the Bailie, with an air of reverence, which I have observed the Scotch usually pay to their distinguished rivers. The Clyde, the Tweed, the Forth, the Spey, are usually named by those who dwell on their banks with a sort of respect and pride, and I have known duels occasioned by any word of disparagement.'—*Rob Roy*, vol. ii. chap. xi.

its slopes may change their aspect, now clothed in bent or purple heather, now waving with bracken or birken copsewood, now striped with fields of tillage, but the clear river that dashes merrily onward through these diversities of scene unites them all into one continuous dale.

The isolation imposed on the separate communities by this topography of the ground inured them to habits of self-dependence. It gave them a coherence that served them in good stead for attack or defence in the old days of Border forays. Each stream not only gave its name to the whole valley which it traversed, but to the human population that dwelt by its banks. It was called a 'Water,' such as Leader Water, Allan Water, Jed Water, and many more, and this word 'water' came to be synonymous with the able-bodied inhabitants of the dale. When, for instance, old Buccleuch gave his orders for the ride to rescue Jamie Telfer's cattle, carried off by English thieves, he bade his men

> Gar warn the water, braid and wide,
> Gar warn it sune and hastilie.

The marauding propensities of one of these communities would sometimes be condensed into the name of their valley, as where Dick o' the Cow complains that

> Liddesdale's been i' my house last night,
> And they hae taen my three kye frae me.

The ballads are so full of human incident as to leave little room even for a background of landscape, but some of the features of the scenery are here and there graphically indicated by a line or even a word. 'The

bent sae brown' of the higher ground gives place to
'heathery hill and birken shaw,' with here and there
a 'bush of broom' or 'buss o' ling' where the dun
deer couches in the glade.   We are led to where

> The hills are high on ilka side
> An' the bought i' the lirk o' the hill.

We are made to see that the 'morning sun is on the
dew,' to feel 'the cauler breeze frae off the fells,' and to
note here and there 'the gryming of a new-fa'n snaw.'
When the king led his army through Caddon ford, in
pursuit of the outlaw Murray, and came in sight of
Ettrick forest, the ballad tells how

> They saw the darke Forest them before,
> They thought it awsome for to see.

But perhaps the natural feature most frequently alluded
to in the tales of foray is the flooding of the rivers.   In
those days bridges were few throughout the Border,
and thus a heavy downfall of rain might completely
sever all communication between the two sides of a
dale.   To plunge into these swollen torrents was sometimes the only escape from pursuit, and required fully
as much courage and nerve as to stay and face the
approaching foe.   In the famous ride to Carlisle for
the rescue of Kinmont Willie, the party found when they
came to the Eden that

> The water was great and meikle of spate.

But they dashed into it, losing neither man nor horse,
but encountering still worse weather on the English
side—

> The wind began fu loud to blaw;
> But 'twas wind and weet and fire and sleet
> When they came beneath the castle wa'.

On their return with their rescued comrade to the river, they saw that it 'flowed frae bank to brim,' but nothing daunted, they plunged into the flood and safely swam across. Their pursuers, however, gave up the chase at sight of the rushing torrent—

> All sore astonish'd stood Lord Scroope
> He stood as still as rock of stane;
> He scarcely dared to trew his eyes,
> When through the water they had gane.
> 'He is either himsell a devil frae hell,
> Or else his mother a witch maun be;
> I wadna hae ridden that wan water
> For a' the gowd in Christentie.'

But even in the midst of the rough warfare of these olden days, there was often a thread of tender affection and romance woven by the ballad-singers into their tales. The vale of the river Yarrow has been more specially consecrated by these tragic songs, and the 'dowie howms o' Yarrow' have come to be identified with all that is most pathetic in the minstrelsy of the Border. From the time of the early ballads a succession of minor poets had sung of this vale, until the pathos of its history was fully revealed to the whole world by Scott and Wordsworth.

It may be readily granted that the fascination of Yarrow has mainly sprung from the recollection of the human incidents which have been transacted there, and which have been enshrined in so much touching verse. But these tragic associations will not, I think, of themselves wholly account for this fascination, nor for the sad tone of the poetry. There seems to me to be a source of peculiarly impressive power in the scenery

of the valley itself, and that to this source not a little
of the glamour of Yarrow is to be attributed.   Nowhere
throughout the whole range of the uplands are their
characteristic aspects more perfectly displayed.   Down
the centre of the dale runs the strip of level green
haugh, through which the stream meanders from side
to side across banks of shingle, with a murmuring
cadence that is borne down on the wind like a low
plaintive wail.   On either hand, the smooth green slopes
rise into the rounded summits of the fells, mottled here
with sheets of bracken and there with folds of heather.
The declivities are indented by little side-valleys, each
leading a clear rivulet between grassy banks to the
main stream.   Nowhere do any rugged features mar
the gentle undulations of the ground.   The outer world
seems to lie far beyond the high hills that enclose and
shelter the quiet valley.   There is a deep silence over
the scene, broken now and then by the melancholy
scream of the curlew or the mournful note of the plover.
The mind, amid such surroundings, easily glides from the
present into reverie amidst the past.   The ruined peel
seems to whisper tales of 'old unhappy far-off things
and battles long ago.'   The greener grass around some
mouldering stones points to hamlets long since for-
saken and forgotten.   The scattered birks and alders
recall the 'fair forest' that once clothed the valley with
'many a seemly tree,' and when in these tracts, now
sacred only to sheep, there were

> Hart and hynd, and dae and rae,
> And of a' wild beasts great plentie.

There is a natural expectation in the mind that

scenery which has made for itself so notable a place in the history of English poetry should present, when first seen, some special charm of attractive beauty. And doubtless many have shared the disappointment so well expressed by Wordsworth and by Washington Irving, who nevertheless had the advantage of being shown over the Border country by its great minstrel himself. But with the instinct of a true poet, Wordsworth soon recovered from his first surprise, and divined the inner spirit of the landscape. Nowhere has that spirit been more felicitously expressed than in his second poem on Yarrow. Contrasting his first anticipation with what he found to be the reality, he addressed the vale :—

> Thou, that didst appear so fair
> To fond imagination,
> Dost rival in the light of day
> Her delicate creation :
> Meek loveliness is round thee spread,
> A softness still and holy ;
> The grace of forest charms decayed
> And pastoral melancholy.

On the influence of the upland scenery of southern Scotland upon the genius of Scott I must not enter. He spent his boyhood within sight of these hills, he made them his chief home throughout life, and when, shattered in health and fortunes, he returned from Italy, it was among these hills, and in hearing of the murmur of the Tweed, that he wished to die. No one can read the 'Lay of the Last Minstrel,' or 'Marmion,' without coming under the spell of the Border scenery. Among the descriptive sketches of landscape in the Waverley Novels, none are more lovingly and graphically painted than those where Scott drew from the vivid recollections of his journeys among the dales and moors of the Southern Uplands.

III. For the purposes of our present inquiry, I would class together as HIGHLANDS all the higher, more rugged, and mountainous ground, which differs on the whole from the uplands, not only in its greater elevation, but in the more irregular form of its surface, the narrower crests of its ridges, and the more peaked shapes of its summits. The geological structure of these tracts of country is generally so complicated that it gives rise to much greater variety of outline than is to be found in either of the other types of scenery. Each kind of rock yields to the weather in its own characteristic way, and as the rainfall is heavier and the slopes steeper than elsewhere, the influence of the weather upon the topography is more especially prominent.

In the northern parts of Wales where a group of ancient volcanic rocks has been laid bare by the stupendous denudation of the surface, a small tract of truly highland scenery has been carved out in Cader Idris, Arenig, Snowdon, and the surrounding heights in Caernarvonshire and Merionethshire. Another isolated area of volcanic hills forms the picturesque district of the Lakes. But it is in Scotland that this type is displayed on the largest scale and in the most varied diversity. The Scottish Highlands are built up of the most ancient rocks of the British Islands, and possess a geological structure of extraordinary complexity. They include a vast variety of materials which, rising to the highest elevations in the country, and exposed to the severest climate, have impressed their individual characters upon the landscapes. Thus, in the north-west, where the simplest grouping of rocks is to be seen, masses of horizontal dark-red sandstone have been carved into the huge pyramids that form so singular a feature in the scenery of Ross and Suther-land. In the central and south-western counties, gleam-ing white cones show where the quartzites rise to the surface. In the eastern Grampians, high craggy moors, encircled with stupendous cones and precipices, mark the sites of the bosses of granite. Among the Western Isles, dark splintered crests and pinnacles point out the position of the gabbros. Perhaps the most rugged ground is to be seen among the mica-schists, which com-bine a wonderful array of pointed peak and notched ridge, with tumultuous masses of craggy declivity.

In the eastern Grampians, the mountains include broad tracts of undulating moorland, which, though lying along the summits of the chain, are level enough to be capable of conversion into racecourses. These hills are separated from the sea by a tract of lowland, and lie thus entirely inland. On the west side of the Highlands, however, the ground between the straths and glens mounts upwards into narrow ridges, not in-frequently sharpened into knife-edged crests, while the whole aspect of the landscape is rugged, rocky, and bare. Land and sea appear there to be inextricably intermingled. Islands, peninsulas, and promontories are penetrated or surrounded by sounds and sea-lochs, in such a curious way that even in what might be thought to be the very heart of the country, the tides of the Atlantic are found to ebb and flow into remote and solitary glens. The mountains plunge abruptly into the salt water, and for the most part only along the valley-bottoms are little strips of level land to be seen.

As a consequence of this intricate interlacing of land and sea, the warm, damp breezes from the Atlantic furnish abundance of mist, cloud, and rain to the western Highlands. Thus to the wildness of rugged mountains and stormy firths, there is added a marvellous range of atmospheric effect. Nowhere in Britain can such an union be beheld of picturesque mountain-form and of clear and vivid colour. Nowhere is the grandeur of a winter storm more impressive than when a south-westerly gale drives the breakers against the headlands, howls up the glens, and fills every gully with a foaming torrent.

The scenery of the western Highlands of Scotland was first brought prominently before the world by the publication in the year 1760 of what purported to be *Fragments of Ancient Poetry, collected in the Highlands.* The success of this volume encouraged the translator, James Macpherson, to prepare a much larger collection which he combined into an epic poem and published in 1762, under the name of ' Fingal.' A second epic ' Temora,' appeared during the following year. Keen discussion arose as to the authenticity of these poems. They were by one group of writers upheld as a price- less contribution to literature, recovered by the skill and labour of one man from the lips of the peasantry, and from faded manuscripts that handed down the traditions of a long vanished past. By another class of disputants they were branded as impudent forgeries palmed off upon the credulity of the world by Macpherson himself.

Into this unhappy and still unsettled controversy I have no intention of entering. For my present purpose it is not necessary to decide whether the so-called poems of Ossian were genuine ancient Celtic productions or were entirely fabricated after the middle of last century, though I think we may safely steer a middle course between the extreme views that have been put forward on either side. Few persons now believe that Macpherson's 'Epics' ever existed as such among the Highlanders. But, on the other hand, it is generally admitted that he really did find a number of Ossianic fragments and that he strung these together, no doubt with copious connecting material of his own. How much was genuine and old, and how much spurious and modern, has never yet been satisfactorily determined. But in estimating the influence of Macpherson's 'Ossian' on literature, we have no need to consider the age of the poems. None of these were known to the world at large until 1760, and we have therefore only to concern ourselves with their history from that year onwards.

Those who have engaged in the controversy have almost wholly entered it from the literary or antiquarian side. I prefer to approach it from the side of the scenery and topography of the West Highlands, and to inquire how far the Ossianic landscape was a true representation of nature, whether there was anything in it new to our literature, and whether it exerted any lasting effect on the attitude of society towards the type of scenery which it depicted.

Macpherson had previously published some English verse of little merit ~~and~~ which attracted no notice. But the appearance of his Ossianic translations at once made him famous. The ' Poems of Ossian ' not only became popular in this country, but were translated into the more important languages of the Continent.

In studying the landscape of Macpherson's 'Ossian' we soon learn that it belongs unmistakably to Western Argyleshire. Its union of mountain, glen, and sea removes it at once from the interior to the coast. Even if it had been more or less inaccurately drawn, its prominence and consistency all through the poems would have been remarkable in the productions of a lad of four-and-twenty, who had spent his youth in the inland region of Badenoch, where the scenery is of another kind. But when we discover that the end-less allusions to topographical features are faithful delineations, which give the very spirit and essence of the scenery, we feel sure that whether they were written in the eighteenth century or in the third, they display a poetic genius of no mean order.

The grandeur and gloom of the Highland mountains, the spectral mists that sweep round the crags, the roar of the torrents, the gleams of sunlight on moor and lake, the wail of the breeze among the cairns of the dead, the unspeakable sadness that seems to brood over the landscape whether the sky be clear or clouded—these features of west Highland scenery were first revealed by Macpherson to the modern world. This revelation quickened the change of feeling, already begun, in regard to the prevailing horror of mountain-scenery. It brought before men's eyes some of the fascination of the mountain-world, more especially in regard to the atmospheric effects that play so large a part in its landscape. It showed the titanic forces of storm and tempest in full activity. And yet there ran through all the poems a vein of infinite melancholy. The pathos of life manifested itself everywhere, now in the tenderness of unavailing devotion, now in the courage of hopeless despair.

'Ossian' fascinated some of the greatest men of the time. These Celtic poems, in the words of Mathew Arnold, passed 'like a flood of lava through Europe.' In the deliberate judgement of this acute critic, they revealed 'the very soul of the Celtic genius, and have the proud distinction of having brought this soul of the Celtic genius into contact with the genius of the nations of modern Europe, and enriched all our poetry by it [1].' There can at least be no doubt that they gave a new and powerful impulse to the appreciation of the wilder aspects of nature, and did much to prepare the way for that love of mountain-scenery which has been one of the characteristic developments of the present century. It is not that in Ossian Highland landscape was deliberately described, but it formed a continually visible and changing background. The prevalent character of the whole range of scenery in the region, and the general impression made by it on the eye and mind, were so vividly conveyed that no one familiar with the country

---

[1] Arnold, *On the Study of Celtic Literature*, 1867, p. 152.

can fail to recognize how faithfully the innermost spirit
of the West Highlands is rendered.

Never before or since have the endless changes of
sky and atmosphere been more powerfully portrayed.
In the tempestuous climate of the west of Scotland
these changes succeed each other with a rapidity and
energy such as the dweller on the southern lowlands
can hardly realize. They are faithfully, if somewhat
monotonously, reflected in 'Ossian.' All through the
poems the air seems ever astir around us. Sometimes
it is only a gently-breathing zephyr which

> Chases round and round
> The hoary beard of thistle old,
> Dark-moving over grassy mounds[1].

We mark the graves of dead heroes by

> Their long grass waving in the wind,

and we move along 'in the robe of the misty glen'
past

> Branches and brown tufts of grass
> Which tremble and whistle in the breeze.

---

[1] The quotations here given are from Dr. Clerk's translation of
Macpherson's Gaelic version of the Poems. The question has
been much disputed whether his English or Gaelic is the original.
There can be no doubt that on the whole the Gaelic is more vivid
and accurate in the description of landscape than the more vague
and bombastic English of Macpherson. Dr. Clerk, who has given
a literal rendering of the Gaelic line for line, remarks :—' I believe
that a careful analysis would resolve very much of Ossian's most
weird imagery into idealized representations of the ever-varying
and truly wonderful aspects of cloud and mist, of sea and mountain, which may be seen by every observant eye in the Highlands,
and it is no fancy to say that the perusal of these poems, as we
have them, may be well illustrated by travelling a range of the
Highland mountains.'—*Poems of Ossian*, Dissertation, vol. i. p. lxv.

But when the full Atlantic gale sweeps over the land, and the rain-clouds rush in swift procession across the half-hidden hills, the moaning and shrieking of the storm come like sounds from another world. We seem to hear the tread, and almost to see the forms, of the ghosts of the Ossianic heroes,

> Chasing spectre-boars of mist
> On wings of great winds on the cairn.

> When bursts the cloud in Cona of the glens,
> A thousand spirits wildly shriek
> On the waste wind that sweeps around the cairn.

Nor is the turmoil of the tempest on the sea less vividly depicted. We are shown the

> Waves surging onward in mist,
> When their crests are seen in foam
> Over smoke and haze widespread.

In the midst of the gloom we descry a shore-stack against which the ocean

> Dashes the force of billows cold;
> White spray is high around its throat,
> And cairns resound on the heathery steep.

With these pictures of tumult on land and sea, there come glimpses of those cherished interludes of bright sunshine, when the western hills and firths are seen at their loveliest. But whether radiant or gloomy the landscape is in unison with the human emotion described—

> Pleasing the tale of the time which has gone;
> Soothing as noiseless dew of morning mild,
> On the brake and knoll of roes,
> When slowly rises the sun

> On the silent flank of hoary Bens—
> The loch, unruffled, far away,
> Calm and blue on the floor of the glens[1].

As a final sample of the Ossianic landscape, with its kaleidoscopic play of atmospheric effect, answering to the changes of human feeling, let me cite some lines from ' Fingal ' :—

> Morna, most lovely among women,
> Why by thyself in the circle of stones,
> In hollow of the rock on the hill alone?
> Rivers are sounding around thee;
> The aged tree is moaning in the wind;
> Turmoil is on yonder loch;
> Clouds darken round the tops of Cairns [mountains];
> Thyself art like snow on the hill—
> Thy waving hair like mist of Cromla,
> Curling upward on the Ben,
> 'Neath gleaming of the sun from the west;
> Thy soft bosom like the white rock
> On bank of Brano of white streams[2].

Though Macpherson roused the interest of the world in the rugged scenery and boisterous climates of the west, it was some time before any other writer followed his lead among the highlands of this country. It is singular to reflect that though the mountain-world, more than any other part of the land, appeals to the imagination, by revealing all that is most impressive in form and colour, and all that is most vigorous in the elemental warfare of nature, it was the last part of the terrestrial surface to meet with due appreciation. Little more than a century has passed since men began to visit the Scottish Highlands for the pleasure of

---

[1] ' Fingal,' iii. 3.   [2] Ibid. i. 211.

admiring their scenery. Previous to the suppression of the Jacobite rising, that mountainous region was regarded as the abode of a half-savage race, into whose wilds few lowlanders would venture without the most urgent reasons. Even after military roads were made across it, the accommodation for travellers was generally of the most wretched kind. Those who had occasion to traverse it gave such an account of their experiences as one would hardly now expect to receive from the heart of Africa [1]. The poet Gray during his visit to Scotland in the year 1765, made a brief excursion into the Perthshire Highlands, and, in spite of the discomforts of travel at that time, came away with a vivid impression of the grandeur and beauty of the scenery. But the only record that remains of this impression is to be found in a few sentences in his letters [2].

---

[1] In Burt's *Letters*, which give so graphic a picture of the condition of the Highlands of Scotland between the two risings of 1715 and 1745, the general impression made at that time on the mind of an intelligent stranger by the scenery of the region may be gathered from the following quotations :—' I shall soon conclude this description of the outward appearance of the mountains, which I am already tired of, as a disagreeable subject. . . . There is not much variety, but gloomy spaces, different rocks, heath, and high and low, . . . the whole of a dismal gloomy brown drawing upon a dirty purple; and most of all disagreeable when the heath is in bloom. But of all the views, I think the most horrid is, to look at the hills from east to west, or vice versa ; for then the eye penetrates far among them, and sees more particularly their stupendous bulk, frightful irregularity, and horrid gloom, made yet more sombrous by the shades and faint reflections they communicate one to another.'—*Letters from a Gentleman in the North of Scotland to his Friend in London.* Fifth edit., vol. i. p. 285.

[2] In writing to Mason he says : ' I am returned from Scotland,

Eight years after Gray's visit, Samuel Johnson in
1773 made his more adventurous journey to the
Hebrides. When we consider what were the discomforts, and sometimes the actual dangers which he had
to undergo, we cannot but admire the quiet courage
with which he endured them, and the reticence with
which he refers to them in his narrative. But Johnson
could see no charm in the Highland mountains. In his
poem on London he had asked many years before :—

> For who would leave, unbrib'd, Hibernia's land,
> Or change the rocks of Scotland for the Strand?

Yet when at last he set foot in Scotland, he showed
no disposition to prefer its rocks to his haunts in
London. Travelling through some of the finest scenery
in Western Inverness-shire, this is the language he
uses regarding it: 'The hills exhibit very little variety;
being almost wholly covered with dark heath, and even
that seems to be checked in its growth. What is not
heath is nakedness, a little diversified by now and then
a stream rushing down the steep. An eye accustomed

---

charmed with my expedition: it is of the Highlands I speak; the
Lowlands are worth seeing once, but the mountains are ecstatic,
and ought to be visited in pilgrimage once a year. None but
those monstrous creatures of God know how to join so much
beauty with so much horror. A fig for your poets, painters,
gardeners, and clergymen, that have not been among them, their
imagination can be made up of nothing but bowling-greens,
flowering shrubs, horse-ponds, Fleet ditches, shell-grottoes, and
Chinese rails. Then I had so beautiful an autumn; Italy could
hardly produce a nobler scene, and this so sweetly contrasted with
that perfection of nastiness and total want of accommodation, that
Scotland only can supply.'—Gray's *Works*, edit. E. Gosse, vol. iii.
p. 223.

to flowery pastures and waving harvests is astonished and repelled by this wide extent of hopeless sterility. The appearance is that of matter incapable of form or usefulness, dismissed by nature from her care and disinherited of her favours, left in its original elemental state, or quickened only by one sullen power of useless vegetation [1].'

While such was the attitude of the man of letters in this country, influences were at work on the continent which powerfully affected the relations of literature to the whole realm of outer nature, and more especially to mountain-scenery. Rousseau's descriptions, followed by the more detailed and scientific narrative of De Saussure, drew the attention of society to the fascinations of Switzerland and the Alps. But these influences had hardly had time to exert much sway in their application to the scenery of our own country when the genius of Scott suddenly brought the features of the Scottish Highlands into the most popular literature of his day. In his youth the future poet and novelist had paid frequent visits to the glens and lakes of Perthshire, where he found many a primitive custom still remaining, which has since vanished before roads, railways, and tourists. In the year 1810 his ' Lady of the Lake' appeared. Thenceforward the stream of summer visitors set in, which has poured in an ever-increasing flood into the Highlands of Scotland. The general interest thus awakened in the glens and mountains of the north was still further intensified by the advent of

---

[1] *Journey to the Western Islands of Scotland,* 1775, p. 84.

the 'Lord of the Isles,' and of 'Waverley,' 'Rob Roy,' and the other novels that depict scenes in the Highlands. Certainly no man ever did so much as Walter Scott to make the natural features of his native country familiar to the whole world. The literary charm which he threw over the hills and glens of Perthshire kindled a wide-spread enthusiasm for the more rugged aspects of nature, and gave a powerful stimulus to the slowly-growing appreciation of the beauty and grandeur of mountain-scenery.

Nevertheless it must be admitted that Scott's high-land landscapes, though more prominent and detailed than those in his descriptions of the lowlands and uplands, were also more laboured and less sponta-neous. His pictures are no doubt faithful and graphic, and each of them leaves on the mind a clear impression of the scene depicted. But their effect is produced rather by a multiplicity of touches than by a few master-strokes of poetic insight and graphic delineation. More-over they are all in one tone of colour, and lack that changeful diversity so characteristic of mountains. They are chiefly fine-weather portraits, as if the poet loved only summer sunshine among the hills, and had either never seen or cared not to portray their gloom, cloud, and storm. We are bound of course to remember that, after all, he was only an occasional visitor to the High-lands. He had not been born among them, and never lived long enough in their solitudes to become inti-mately versed in all their alternations of mood under changes of sky and season. He writes of them as an admiring and even enthusiastic spectator, but not as one into whose very soul the power of the mountains had entered. He never warms among them into that fervent glow of affectionate appreciation which kindles within him in sight of the landscapes of his native Border.

One other mountainous district in Britain—that of the English Lakes—claims our attention for its influence on the progress of the national literature. Of all the isolated tracts of higher ground in these islands, that of the Lake District is the most eminently highland in character. It is divisible into two entirely distinct portions by a line drawn in a north-easterly direction from Duddon Sands to Shap Fells. South of that line the hills are comparatively low and featureless, though they enclose the largest of the lakes. They are there built up of ancient sedimentary strata, like those that form so much of the similar scenery in the uplands of Wales and the south of Scotland. But to the north of the line most of the rocks are of a different nature, and have given rise to a totally distinct character of landscape. They consist of various volcanic materials which in early Palaeozoic time were piled up around submarine vents and accumulated over the sea-floor to a thickness of many thousand feet. They were subsequently buried under the sediments that lie to the south, but, in after ages uplifted into land, their now diversified topography has been carved out of them by the meteoric agents of denudation. Thus pike and fell, crag and scar, mere and dale, owe their several forms to the varied degrees of resistance to the general waste offered by the ancient lavas and ashes. The upheaval of the district seems to have produced a dome-shaped elevation, culminating in a summit that lay somewhere between Helvellyn and Grasmere. At least from that centre the several dales diverge, like the ribs from the top of a half-opened umbrella.

The mountainous tract of the Lakes, though it measures only some thirty-two miles from west to east by twenty-three from north to south, rises to heights of more than 3,000 feet, and as it springs almost directly from the margin of the Irish Sea, it loses none of the full effect of its elevation. Its fells present a thoroughly highland type of scenery, and have much of the dignity of far loftier mountains. Their sky-line often displays notched crests and rocky peaks, while their craggy sides have been carved into dark cliff-girt recesses, often filled with tarns, and into precipitous scars, which send long trails of purple scree down the grassy slopes.

Moreover, a mild climate and copious rainfall have tempered this natural asperity of surface by spreading a greener mantle over the lower parts of the fells and the bottoms of the dales than is to be seen among the mountains further north. Though the naked rock abundantly shows itself, it has been so widely draped with herbage and woodland as to combine the luxuriance of the lowlands with the near neighbourhood of bare cliff and craggy scar.

Such was the scenery amidst which William Wordsworth was born and spent most of his long life. Thence did he draw the inspiration which has done so much to quicken the English poetry of this century, and which has given to his dales and hills so cherished a place in our literature. The scenes familiar to him from infancy were loved by him to the end with an ardent and grateful affection which he never wearied of publishing to the world. No mountain-landscapes had ever before been drawn so fully, so accurately, and in such felicitous language. Every lineament of his hills and dales is depicted as luminously and faithfully in his verse as it is reflected on the placid surface of his beloved meres, but suffused by him with an ethereal glow of human sympathy. He drew from his mountain-landscape everything that

> Can give an inward help, can purify
> And elevate, and harmonize and soothe.

It brought to him 'authentic tidings of invisible things'; filled him with

> The sense
> Of majesty and beauty and repose,
> A blended holiness of earth and sky.

For his obligations to that native scenery he found continual expression.

> Ye mountains and ye lakes,
> And sounding cataracts, ye mists and winds
> That dwell among the hills where I was born,
> If in my youth I have been pure in heart,
> If, mingling with the world, I am content
> With my own modest pleasures, and have lived
> With God and Nature communing, removed
> From little enmities and low desires—
> The gift is yours.

Not only did his observant eye catch each variety of form, each passing tint of colour on his hills and valleys, he felt, as no poet before his time had done, the might and majesty of the forces by which, in the mountain-world, we are shown how the surface of the world is continually modified.

> To him was given
> Full many a glimpse of Nature's processes
> Upon the exalted hills.

The thought of these glimpses led to one of the noblest outbursts in the whole range of his poetry, where he gives way to the exuberance of his delight in feeling himself, to use Byron's expression, 'a portion of the tempest'—

> To roam at large among unpeopled glens
> And mountainous retirements, only trod
> By devious footsteps; regions consecrate
> To oldest time; and reckless of the storm,
> .   .   .   .   . while the mists
> Flying, and rainy vapours, call out shapes
> And phantoms from the crags and solid earth,
> .   .   .   .   . and while the streams
> Descending from the region of the clouds,
> And starting from the hollows of the earth,
> More multitudinous every moment, rend
> Their way before them—what a joy to roam
> An equal amongst mightiest energies!

In this passage Wordsworth seems to have had what he would have called 'a foretaste, a dim earnest' of that marvellous enlargement of the charm and interest of scenery due to the progress of modern science. When he speaks of 'regions consecrate to oldest time,' he had a vague feeling that somehow his glens and mountains belonged to a hoary antiquity, such as could be claimed by none of the verdant plains around. Had he written half a century later he would have enjoyed a clearer perception of the vastness of that antiquity

and of the long succession of events with which it was
crowded [1].

It is curious to remember that three of the poets
whom I have singled out as illustrations of the influence
of our lowland, upland, and highland scenery upon our
literature have held up the geologist to ridicule. Cowper
put that votary of science into the pillory among the
irreligious crowd, about whose ears the poet loved to
'crack the satiric thong [2].' Wordsworth treated the
geological enthusiast with withering scorn [3]. Scott, in

---

[1] Sedgwick did his best to enlighten the poet by his famous
Four Letters on the Geology of the Lake District; but these came too
late. They were published at Kendal in 1846, and Wordsworth
died in 1850.

[2]   Some drill and bore
   The solid earth, and from the strata there
   Extract a register, by which we learn,
   That He who made it, and revealed its date
   To Moses, was mistaken in its age.—'The Task,' bk. iii. 150.

[3]   You may trace him oft
   By scars which his activity has left
   Beside our roads and pathways, though, thank Heaven!
   This covert nook reports not of his hand—
   He who with pocket-hammer smites the edge
   Of luckless rock or prominent stone, disguised
   In weather-stains or crusted o'er by Nature
   With her first growths, detaching by the stroke
   A chip or splinter—to resolve his doubts;
   And, with that ready answer satisfied,
   The substance classes by some barbarous name,
   And hurries on; or from the fragments picks
   His specimen, if but haply interveined
   With sparkling mineral, or should crystal cube
   Lurk in its cells—and thinks himself enriched,
   Wealthier, and doubtless wiser, than before!
                              'The Excursion,' bk. iii.

his characteristic good humour, only poked fun at him[1]. It was reserved for a poet of our own day to look below the technical jargon of the schools, and to descry something of this wealth of new interest which the landscape derives from a knowledge of the history of its several parts. But Tennyson only entered a little way into this enlarged conception of nature. There remains a boundless field for some future poetic seer, who letting his vision pierce into the past, will set before the eyes of men the inner meaning of mountain and glen.

And thus, while we recognize the potent influence which the scenery of the country has exerted on the progress of our literature, we can look forward to a fresh extension of this influence as the outcome of geological investigation. Already the result of this widening of the outlook has made itself felt alike in prose and verse. The terrestrial revolutions of which each hill and dale is a witness; the contrasts presented between the present aspect and past history of every crag and peak; the slow silent sculpturing that has carved out all this marvellous array of mountain-forms— appeal vividly to the imagination, and furnish themes that well deserve poetic treatment. That they will be seized upon by some Wordsworth of the future, I cannot doubt. The bond between landscape and literature will thus be drawn closer than ever. Men

---

[1] 'Some rin up hill and down dale, knapping the chucky stanes to pieces wi' hammers, like sae mony roadmakers run daft—they say it is to see how the warld was made.'—Meg Dods *loq.* in 'St. Ronan's Well,' chap. ii.

will be taught that beneath and behind all the outward beauty of our lowlands, our uplands, and our highlands there lies an inner history which, when revealed, will give to that beauty a fuller significance and a new charm.